The Nature and Science of

RAIN

Jane Burton and Kim Taylor

W
FRANKLIN WATTS
NEW YORK • LONDON • SYDNEY

First published in 1997

Franklin Watts
96 Leonard Street
London EC2A 4RH

Franklin Watts Australia
14 Mars Road
Lane Cove
NSW 2066

Conceived, designed and produced by
White Cottage Children's Books
29 Lancaster Park
Richmond, Surrey TW10 6AB, England

Editor/Art Director: Treld Pelkey Bicknell

Educational Consultant: Jane Weaver

Scientific Advisor: Dr Jan Taylor

Set in Rockwell Light by R & B Creative Services

Originated by R & B Creative Services

Printed in Belgium

ISBN: 0 7496 2924 X

Dewey Decimal Classification Number: 551.57

A CIP catalogue record for this book is available
from the British Library

Contents

Miraculous Rain

Heavy rain can be exciting and even dangerous. Raindrops come pelting down, the ground gets sodden and puddles form. Swirling brown water fills the rivers and may burst over their banks. But after rain, when the sun shines again, leaves are fresh and green, birds sing and flowers open.

Rain is absolutely **vital** to life on Earth, but it is possible only because air and water are able to mix. It may seem unlikely that a gas and a liquid can mix, but that is what happens when water **evaporates** into the air and that is where rain begins.

Water itself is made up of two gases, **hydrogen** and **oxygen**, joined together. Two **atoms** of hydrogen **combine** with one atom of oxygen to make one **molecule** of water. That is why the **chemical formula** for water is H_2O.

A raindrop falls into a still pool causing a spike of water to shoot up and break into smaller drops. ▼

◄ Black clouds have built up along the coastline of west Wales, blotting out the sun. Rain is falling in grey curtains over the sea.

Wet Air

The brilliant colours ▲ of Harlequin Bugs gleam amongst the **moist** green leaves of the **tropical** forest.

All around you there is water in the air. You cannot see the water because it is in the form of separate molecules floating about, mixed in with the molecules of air. It is water in a gas-like form and is known as water **vapour**.

There is always some water vapour in the air. In desert areas there is just a little, but in **rainforests** there is a lot. The amount of water vapour in the air is measured as a percentage called **relative humidity**. When the relative humidity is 100 per cent, it means that the air cannot carry any more water. It is **saturated**, and wet things will not dry at all.

A Green Iguana gets ▲ its water through eating leaves which contain water and through licking raindrops off the foliage.

◄

Plants take water out of the ground and give out water vapour into the air through their leaves. Where there are lots of leaves, as in this rainforest, the air is always moist.

Fog and Clouds

Have you have ever walked down by a river on a misty morning, or been in a city fog, or climbed a mountain until you reach the clouds? If you have, you will have seen the tiny specks of water that make up fog and cloud **drifting** in the air. These tiny water **droplets** are much easier to see in bright light, especially when the sun shines.

Warm air can carry more water vapour than cold air. This means that when air is cooled, its relative humidity rises and it may become saturated if it is cooled enough. Fog and clouds form when saturated air is cooled further. Then, the vapour **condenses** into water droplets. This happens when moist air near the ground rises high into the cooler regions of the **atmosphere** and forms clouds. It also happens at night when the ground becomes cold and cools the air next to it, forming fog.

◀

On a damp, misty autumn morning the air is saturated with water vapour and fine droplets hang in the air. Every surface is wet including the twigs and leaves of the trees, and the only sound is a steady drip, dripping.

Mist forms close to the ground when the air cannot carry any more water vapour. Some of it has to condense and every surface becomes laden with drops – even a piece of thistledown caught in the fine gossamer of a spider's web. ▼

Clouds form when wind carries moisture-laden air over mountains. The wet air is pushed up and cooled, causing the vapour to condense into tiny droplets. ▶

The shallow water where the Greater Flamingos are paddling has been warmed by the sun all day, but now it is evening and the incoming tide is so cold that it is cooling the air next to it, causing mist to form. ▼

Most of the clouds that you see in the sky are made up of millions of minute water droplets **suspended** in the air. All this water is collected by the air in the form of vapour. A lot of it comes from the surface of the sea – particularly from warm **tropical** seas because warm water **vaporises** more readily than cold water. When salt water evaporates, only the water gets into the air. The salt is left behind.

Cold seas also supply water vapour to the air – especially when whipped up by stormy winds. A cold stormy sea like this one off the coast of Namibia (below) can put a lot of water vapour into the air.

Falling Rain

What causes a cloud to turn into rain? There is an easy answer: it is just a matter of **drop size**. When the droplets in a cloud grow beyond a certain size, instead of drifting, they start to fall.

The reason why droplets in clouds sometimes grow is more complicated, but it is mostly a matter of temperature. When a cloud becomes colder, more water vapour condenses onto its droplets so that they get bigger and start to drift downwards. The falling droplets gather more water on their way, soon becoming raindrops which tumble towards the Earth.

Raindrops falling ▶ from a thunder cloud collect moisture on their way down. The further they have to fall, the bigger they get and the faster they fall. They hit the ground really hard when they get there, and puddles quickly form. The black clouds are reflected on the surface of the puddles, and everywhere is dark.

Clouds become cooled if the air that carries them is pushed up higher into the atmosphere. This happens when wind blows them over mountains. That is why mountainous areas close to the sea are some of the wettest places on Earth. On high mountains, the air is so cold that the wetness falls as snow.

Thunder storms ▲ often form when **currents** of warm moist air rise high in the sky above ground that has been heated by the sun.

Drips and Drops

The water droplets in clouds and fine raindrops drifting gently downwards are perfect **spheres**. They are like tiny round beads. Each drop is held in shape by something called **surface tension**. This is like an elastic skin which keeps each drop round. All water surfaces—even the flat surfaces of lakes and puddles—have this invisible elastic **film** over them.

As falling raindrops get bigger, they hurtle through the air faster and their shape changes. The rush of air pushing against their lower sides makes them egg shaped. Eventually, surface

▲ A drop falls off a berry... ▼

▼ It hits the surface of a pond...

▼ The drop smacks into the pond making a hollow in its surface...

▼ Water rushes from below to fill the hollow, but goes too far and makes a spike.

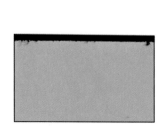

Raindrops falling into a pond hit the surface so hard they make little spikes jump up out of the water. You can see a spike with a drop above it in between the ducks.

tension, which is not very strong, can no longer hold each drop together and the big drops break up into smaller drops. This means that there is a **maximum** size for raindrops beyond which they cannot grow any bigger. The largest raindrops are only just over 4 millimetres across.

▼A little drop breaks off the top of the spike...

▼ and keeps going up.

▼ The spike falls back into the water...

▼ leaving the little drop to fall in later.

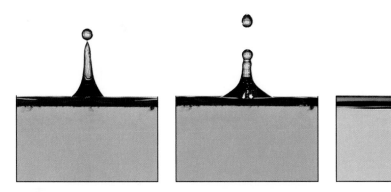
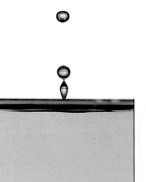
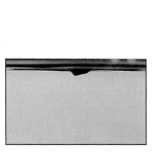

Looking Through Raindrops

A raindrop hanging on a twig is a **lens**. Look into it and you can see a tiny upside-down **image** of what is beyond. A lens **focuses** light so that when you look through it, things look bigger or smaller, depending on how far the lens is from what you are looking at.

When the sun shines, raindrops hanging in bushes and trees also act as **prisms** and glitter with rainbow colours. One drop sparkles a brilliant blue, another sparkles red, another green. If you move your head slightly, each drop will change colour. This happens because sunlight is a mixture of all the colours of the rainbow. Each raindrop separates the colours so that they shine out at slightly different angles. The colour you see depends on the angle between the sun, the raindrop and you – and when you move your head, you change the angle slightly.

A raindrop caught in a Nasturtium leaf makes a large, right-way-up image of the veins in the leaf.

A raindrop hangs from a pansy bud. Look carefully and you can see in it an upside-down image of the two pansies beyond. The raindrop is acting as a lens.

Raindrops hanging on leaves and twigs glitter with brilliant colours in sunlight. Each drop is a tiny prism that splits white light from the sun into rainbow colours.

Waterproofing

Rain is possible because air and water are able to mix. But **oil** and water on their own cannot mix, and many animals use oil for **waterproofing**. Keeping dry is important for most animals, and oily skins save many small creatures from being drowned by rain.

Oil is used by birds for waterproofing their feathers. Most birds have a special oil **gland**

You can just see this Barn Owl's oil gland— a little pink spike just above the tail. The owl has taken some oil with its beak and is smearing it onto its tail feathers to waterproof them as it **preens**.

▶

above the tail. When a bird preens, it takes oil from the gland with its beak and smears the oil over its feathers. This thin film of oil makes raindrops bounce off feathers without wetting them.

Furry mammals also stay dry-skinned in the rain. Oil glands in the skin keep their coats sleek and waterproof.

As well as animals, many land plants need to keep dry. If rainwater gets into plant leaves and stems, it **dilutes** the **sap** and damages the **cells** so that the plant cannot grow properly. To keep rainwater out, many plants have a thin layer of **wax** on their outsides. The wax also helps to stop the plant losing water from its sap during dry weather.

These rats have just had a swim and water drips off their fur. Oil from glands in the skin keeps the fur waterproof.

▼ In dry weather a Wood-sorrel leaf is wide open....

▼ Too much rain can damage the delicate leaf and so when the rain starts, the leaf begins to close....

▼ Heavy rain cannot harm the closed leaf.

Rain to Drink

In the rainy season ▲ the Banjo Frog comes out of hiding and sits in a puddle shouting "Boink!" to attract a mate.

All plants and animals have water in their bodies. Plants suck up rainwater through their roots and animals drink from rain pools when they are thirsty. Some soft-bodied animals such as frogs, snails and slugs do not need to drink because they are able to take in water through their skins. But because their skins are **permeable** to water, these animals can only be active in damp conditions. If they come out when the weather is dry, their bodies quickly dry up and **shrivel**.

Very few living things can survive for long without water. But some plant seeds can remain dry for years and yet start to grow when dampened by rain. The eggs of Fairy Shrimps may blow about in desert dust for up to twenty years and yet still be able to hatch into little shrimps if they happen to end up in a rain pool.

This Garden Snail ▲ is taking in water through its skin. But water can also evaporate through its skin and so, in dry weather, the snail seals itself into its shell with slime to stop its body drying out.

Rain is often scarce on the plains of Africa, but animals can smell rain from miles away. These Common Zebras, Blue Wildebeests and Springboks may have come a long way to find this rain pool left after a heavy overnight shower. ▼

Water at Work

Rain may fall as fine **drizzle** or as a steady downpour of big drops. It may be just a shower, lasting only a few minutes, or it may go on for days. Wherever rain falls over land, it mostly drains into streams and rivers which take it back to the sea.

Much of the Earth's surface has been shaped by rainwater. Rivers, swollen by rain, carve deep gorges down mountainsides. Over level ground, they spread out and **meander**, making wide valleys. At the same time, rivers are carrying silt down to the sea. Shaping the land is hard work. A lot of energy is needed to move all this earth.

Rain also cleans the atmosphere, washing out dust and smoke **particles**. Occasionally, it washes so much **pollution** out of the atmosphere that it falls as **acid rain** which damages plants and kills fish.

▲
Rivers shape giant rocks and they shape small pebbles by grinding one against the other. Bits get chipped off and surfaces smoothed. This water-worn wood, seed-heads and bleached body of a dung beetle have been left behind by the floodwaters.

◄
When heavy rain falls, rivers like the Tana in northern Kenya become brown and swirling. Huge quantities of earth are carried away in the water and dumped further downstream or even into the sea.

Rainy Seasons

A dwarf cactus ▲ can survive desert **droughts** by storing water in its round stem.

Rain is so important to plants and animals that their lives are often controlled by the rainy seasons. In many warmer parts of the world, the coming of the rainy season is like spring in cooler parts of the world. Seeds start to grow, plants send out new leaves, birds sing and build their nests and insects emerge in millions.

Not all parts of the world have a regular rainy season. Rain in the desert can be very **irregular**, with some places having no rain at all for years. Only a few very special plants and animals can survive there.

Rain is falling on this baby Warthog who does not look very happy! The rain is cold because it has come down from high clouds where the temperature is near to freezing. But rain means plenty of food for Warthogs, and this baby will soon dry off when the sun comes out again. ▶

Other parts of the world receive rain at all seasons. In tropical rainforests, it may rain almost every day—usually in the afternoon. Here, hundreds of different types of huge trees grow together, draped with vines, ferns, mosses and orchids, and very many kinds of colourful birds and insects live amongst the garden-like tree tops.

At the start of the ▲ rainy season in southern Africa, rain falls in heavy showers leaving much of the land still dry. These giraffes and Impalas may travel to places where rain has fallen, to find fresh grass and leaves as well as water to drink.

The Water Cycle

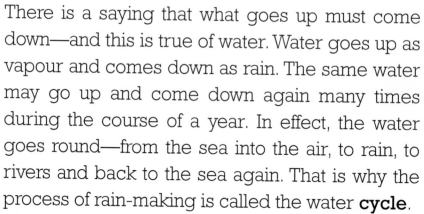

As you watch rain-drops trickling down the window, it is interesting to imagine that the same water, perhaps only one or two days ago, was part of the sea. Fishes and even whales were swimming amongst its molecules!

There is a saying that what goes up must come down—and this is true of water. Water goes up as vapour and comes down as rain. The same water may go up and come down again many times during the course of a year. In effect, the water goes round—from the sea into the air, to rain, to rivers and back to the sea again. That is why the process of rain-making is called the water **cycle**.

The water cycle keeps water moving over the surface of the Earth. The sun provides all the energy needed for this work. Its heat evaporates water from the surface of the sea, turning salt water into fresh water. The heat of the sun makes wind which carries the vapour over land, forming clouds and rain. Every living thing on land needs rain because plants cannot live without water, and without plants there would be nothing for animals to eat. Animals also need water to drink, and many of them spend their whole lives in it.

Sunshine in one place and rain in another are parts of the water cycle. Sun and rain at the same time are needed to make a rainbow. Here, a shaft of sunlight coming through a gap in the storm clouds is being separated by falling raindrops into rainbow colours.

Things to Do:

Measuring and Making Rain

Rain may fall in torrents for a few minutes or it may come down as steady drizzle for hours. The effect rain has on the land depends mostly on the amount of water that falls. Rainfall is measured in millimetres. A shower that gives one millimetre of rain would cover a flat surface with water one millimetre deep.

To make your own rain gauge you need a plastic funnel and a straight-sided clear plastic beaker almost the same size as the top of the funnel. You also need a ruler to measure the rainfall.

Put the funnel into the beaker so that it fits closely (below). It will stop rain splashing out of the beaker. It will also

stop the rain that you have collected from evaporating. Place your rain gauge in a quiet spot away from any trees or buildings. Each day at the same time, put the beaker on a level surface, measure the depth of water in it and write down the measurement. Empty the beaker afterwards.

Some days there may be no rain at all, while on others you may record several millimetres. Keep a diary of rainfall to see how it varies from one time of year to another. All over the world people are recording rainfall in this way and from the records, rainfall maps are drawn showing the areas with heavy, medium or light rainfall.

The Water Cycle in a Bowl

If you want to see how fresh rain comes from the salty ocean, you can make the water cycle happen in a bowl! You need a deep glass ovenware bowl about 25 centimetres across, a glass tumbler and some thin polythene cling-film. You will also need a spoonful of salt and some green or blue food dye to make "sea water".

Put some really hot—but not boiling—water into the bowl. There should be enough water so that the tumbler will just stand on the bottom of the bowl without floating. Stir in the salt plus a few drops of dye and stand the tumbler the right way up in the middle of the bowl. Now you have an island surrounded by sea (above right).

Stretch a layer of cling-film over the top of the bowl and smooth it down the sides so that the bowl is sealed. Next, squeeze a little air out by pressing down

leafy plants. To see how leaves put water vapour into the air you need a bottle and a leafy twig from a tree. Bind some tape around the twig so that, when its cut end is pushed to the bottom of the bottle, the tape makes it fit the neck of the bottle closely. Fill the bottle with water and push in the twig so that its cut end reaches the bottom. Put your experiment in a safe place that is both light and airy. Each day have a look to see how much water the leaves have sucked up and turned into vapour. If you know how much water disappears from the bottle each day and how many leaves there are on your twig, you can calculate how much water per day each leaf is breathing out. Multiply this by a million—because there may be a million leaves on a big tree—and you will know how much water a tree puts into the atmosphere.

A parrot licks moisture from big forest leaves, while the leaves breathe more moisture into the air. ▼

gently on the film while raising a corner of it for a moment to let out the air. This should produce a dish-shaped cover over the bowl which, from underneath, is like a bulging cloud over the island. All you have to do now is to wait and watch.

Water is evaporating from the warm "sea". When the vapour reaches the cool cling-film, it condenses on it as tiny drops. These slowly get bigger until they join together, forming a drip which falls into the tumbler. Rain is falling on the "island"! When the rain is dripping steadily, try pouring some ice-cold water on top of the film and watch what happens. Raindrops should form more quickly. You can see that the rain is not coloured with the dye and later you can check by tasting the water in the tumbler that the rain is fresh water, not salty.

Leaves Make Rain

Water vapour in the air does not only come from the sea. It also comes from the land—particularly land that is covered by

Glossary

Acid rain: Rain containing high levels of acid pollution from power stations and car exhausts.
Atom: The tiny building blocks from which all substances are made.
Atmosphere: The layer of air and clouds that surrounds the Earth

Cells: The microscopic building blocks of plant and animal bodies
Chemical formula: A kind of shorthand used to identify chemicals.
Combine: Join together chemically
Condense: Change from vapour to liquid.
Current: Mass of air or water moving in one direction.
Cycle: Something that goes round in a circle.

Dilute: To add more water.
Drift: To move slowly with a current of air or water.
Drizzle: Very fine rain
Droplet: A tiny drop.
Drop size: The size of the drops in a cloud or in a shower of rain.
Drought: A long time without any rain.

Evaporate: Change from liquid to vapour.

Film: A very thin layer.
Focus: To collect rays of light together to form an image.

Genus: The name given to a group of similar species. Plural: genera. The Large White Butterfly (*Pieris brassicae*) and the Small White Butterfly (*Pieris rapae*) are separate species in the same genus.
Gland: A part of an animal's body that produces a special substance.

Humid: Containing a lot of water vapour.
Hydrogen: A gas. Chemical symbol: H.

Image: A picture.
Irregular: At widely different intervals of time; not regular.

Lens: A piece of clear material with curved sides used to focus light.

Maximum: The largest possible.
Meander: To wander from side to side.
Moist: Containing water. Feeling wet.
Molecule: The smallest part of a substance, made by joining atoms together.

Oil: Kind of liquid that does not mix with water.
Oxygen: A gas forming about one-fifth of the air. Chemical symbol: O.

Particle: A tiny piece of solid material.
Permeable: Allowing water to soak through.
Preen: To smooth feathers and arrange them tidily.

Prism: A block of glass or other clear material used to separate light into rainbow colours.
Pollution: Harmful chemicals in the air or in water.

Rainforest: Forest that grows where there is a lot of rain—usually in tropical regions.
Relative humidity: A measure of the amount of water vapour in the air.

Sap: The liquid found inside plants.
Saturated: Carrying the maximum possible amount of water or water vapour.
Shrivel: To get smaller and become wrinkled.
Silt: Fine particles in or deposited from water.
Species: A biologically distinct kind of animal or plant. Similar species are grouped into the same genus. The word species can be singular or plural.
Sphere: Round like a ball.
Surface tension: The layer of molecules at the surface of a liquid that act like a stretched elastic skin.
Suspend: To hang in air or water.

Tropical: Coming from the warm regions around the Equator.

Vapour: A gas formed from a liquid.
Vaporise: To turn into gas.
Vital: Very important. A matter of life or death.

Wax: A solid oily substance.
Waterproof: Able to keep out water.

Plants and Animals

The *common names* of plants and animals vary from place to place. Their *scientific names*, based on Greek or Latin words, are the same the world over. Each kind of plant or animal has two scientific names—like a first name and a surname for a person—except that the names are placed the other way round. The name of the **genus**, or *generic name*, which is like a surname, always comes first and starts with a capital letter. The name of the **species**, or *specific name*, comes second and always begins with a small letter. In this book, capitals are used for the initial letters of common names to make it clear when a particular species is being referred to.

Hawkbit *(Leontodon* species)—Europe **1**

White-lipped Snail *(Cepaea hortensis)*—Western Europe **3, 5**

Cotoneaster *(Cotoneaster bullatus)*—China; planted elsewhere **5, 14**

Harlequin Bug *(Philia senator)*—Australia **6**

Tree fern *(Cyathea* species)—warm, moist mountainous parts of the world **6-7**

Green Iguana *(Iguana iguana)*—South America **7**

Silver Birch *(Betula pendula)*—Europe, Asia Minor; planted in America **8**

Dog Rose *(Rosa canina)*—Western Europe **9**

Greater Flamingo *(Phoenicopterus ruber)*—Africa, India, Southern Europe **10**

Ginkgo or **Maidenhair Tree** *(Ginkgo biloba)*—China; planted elsewhere **12**

Mallard *(Anas platyrhynchos)*—Europe, North America **15, 18**

Pansy *(Viola tricolor)*—cultivated **16**

Nasturtium *(Tropaeolum majus)*—South America; cultivated worldwide **17**

Barn Owl *(Tyto alba)*—worldwide **18**

Brown Rat *(Rattus norvegicus)*—worldwide **19**

Wood-sorrel *(Oxalis acetocella)*—Europe, Japan **19**

Banjo Frog *(Limnodynastes dorsalis)*—Australia **20**

Common Zebra *(Equus burchelli)*—South and East Africa **20-21**

Blue Wildebeest or **Brindled Gnu** *(Connochaetes taurinus)*—South and East Africa **20-21**

Springbok *(Antidorcas marsupialis)*—Southern Africa **20-21**

Garden Snail *(Helix aspersa)*—Europe **21**

Dwarf cactus *(Rebutia calliantha)*—South America **24**

Warthog *(Phacochoerus aethiopicus)*—Africa **24**

Giraffe *(Giraffa camelopardalis)*—Africa **25**

Impala *(Aepyceros melampus)*—Africa **25**

Yellow-fronted Amazon Parrot *(Amazona ochrocephala)*—South America **29**

Ladies Mantle *(Alchemilla mollis)*—Romania, Asia Minor, planted elsewhere **30**

Tree Mallow *(Lavatera olbia)*—cultivated **31**

American Pillar Rose *(Rosa* species)—cultivated **32**

Index